UNDERSTANDING CLIMATE CHANGE

Extreme Weather and Rising Seas

Karina Hamalainen

Children's Press®
An Imprint of Scholastic Inc.

Content Consultants

Heidi A. Roop, PhD
Research Scientist
Climate Impacts Group
University of Washington, Seattle
Seattle, Washington

Farhana Sultana, PhD
Associate Professor of Geography
Maxwell School of Citizenship & Public Affairs
Syracuse University
Syracuse, New York

Library of Congress Cataloging-in-Publication Data
Names: Hamalainen, Karina, author.
Title: Extreme weather and rising seas: understanding climate change/by Karina Hamalainen.
Other titles: True book.
Description: New York: Children's Press, 2020. | Series: A true book |
Includes index. | Audience: Grades 4-6. (provided by Children's Press.)
Identifiers: LCCN 2019031425| ISBN 9780531130773 (library binding) | ISBN 9780531133774 (paperback)
Subjects: LCSH: Climatic changes—Juvenile literature. | Global warming—Juvenile literature. | Severe
    storms—Juvenile literature. | Global environmental change—Juvenile literature.
Classification: LCC QC941.3 .H35 2020 | DDC 551.6—dc23

Design by THREE DOGS DESIGN LLC
Produced by Spooky Cheetah Press
Editorial development by Mara Grunbaum

All rights reserved. Published in 2020 by Children's Press, an imprint of Scholastic Inc.
Printed in North Mankato, MN, USA   113

SCHOLASTIC, CHILDREN'S PRESS, A TRUE BOOK™, and associated logos are trademarks and/or
registered trademarks of Scholastic Inc.

Scholastic Inc., 557 Broadway, New York, NY 10012

1 2 3 4 5 6 7 8 9 10 R 29 28 27 26 25 24 23 22 21 20

**Front cover: Emergency workers
save a dog from a flash flood.
Back cover: A firefighter battles a blaze.**

Find the Truth!

Everything you are about to read is true *except* for one of the sentences on this page.

Which one is **TRUE**?

T or F Climate change is making hurricanes less wet than they used to be.

T or F The world's oceans are expanding because warm water takes up more space than cold water.

Find the answers in this book.

Contents

Even desert ecosystems will be affected by rising temperatures.

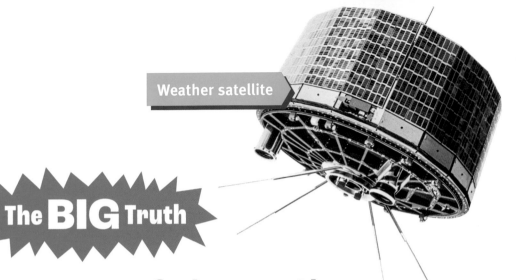

Weather satellite

The BIG Truth

The Costs of Climate Change

Levi Draheim speaks out.

A Critical Moment

In recent centuries, humans have released increasing amounts of **greenhouse gases** into Earth's **atmosphere**. These gases have trapped heat in the atmosphere, causing the average temperature on the planet's surface to rise and contributing to **global warming**.

Because of global warming, oceans are heating up, sea levels are rising, and weather is becoming more extreme. These changes in Earth's climate are known as global **climate change**. They threaten people and other plant and animal species around the world. If we don't make changes to reduce greenhouse gas emissions now, these problems will worsen.

There is good news, though!

Thousands of scientists worldwide are studying global climate change. Politicians, public figures, and citizens of all ages are trying to figure out what to do. Humanity now knows more than ever about the causes and effects of climate change, as well as how we might reduce its impact. That means **people today can make decisions** that will affect the planet for centuries to come.

Turn the page to learn how global warming contributes to worsening storms, floods, and heat waves around the world.

This satellite photo shows two hurricanes moving through the Atlantic at the same time.

HURRICANE

HURRICANE

In June 2019, the Gulf of Mexico experienced its hottest June ocean temperature since record-keeping began 110 years ago!

Heating Up

Earth's climate is always changing. At different times in the planet's history, it has been both hotter and colder than today. So why are people concerned about global climate change now? In the past, Earth's climate changed slowly over thousands of years. However, Earth's surface has warmed by 1.8 degrees Fahrenheit (1 degree Celsius) since just 1880. Even a seemingly small temperature change like this affects sea level, weather patterns, and extreme weather events. These effects are already being seen all around the world.

When a weather forecast is wrong, it's called a "bust."

Even as Earth heats up, snow will continue to fall.

Weather vs. Climate

So if the average temperature is rising around the world, why do many places still get cold and snowy in winter? The key is understanding the difference between climate and weather. Weather is what happens over a short period of time. The weather today might be snowy, or it might be sunny with an afternoon thunderstorm. You can think of weather like a person's mood, which can change from day to day or even hour to hour.

Climate, on the other hand, is the usual weather condition in an area over three decades or more. Scientists analyze the long-term trends and patterns in weather. Over three decades, an area's climate tends to be fairly consistent, like someone's personality. An area with a hot, dry climate is more likely to have warm, sunny weather. An area with a cold, wet climate usually gets more snowy and rainy weather. But rising temperatures around the world are changing what is typical. And as the climate changes, the likelihood of certain weather events does, too.

Warming temperatures are even making it hard for desert plants to survive.

Predicting the Weather

Humans have been observing the weather for thousands of years. People were predicting weather based on cloud patterns at least 3,000 years ago. As time went on, farmers, sailors, and pilots started keeping weather records to do their jobs. But it wasn't until scientists launched the first successful weather **satellites** in the 1960s that we started to get a global picture of Earth's weather and climate, even in hard-to-reach places such as Antarctica.

Timeline: Climate Science

650 BCE: Babylonians study clouds to try to predict the weather.

300 BCE: Chinese astronomers develop a calendar based on seasonal weather changes.

CIRCA 850: Iraqi thinker Al-Kindi studies and writes about the tides, the wind, and weather conditions required for rain, hail, and snow.

1714: German scientist Daniel Fahrenheit invents the first mercury thermometer to measure temperature.

Today, a worldwide network of more than 50,000 weather stations is constantly monitoring the weather. Instruments measure temperature, wind speed, air pressure, precipitation, and more. Scientists feed data from both weather stations and satellites into computers, where software creates weather forecasts. Scientists also use these data to study the climate and how it is changing over time.

1861:
British naval officer Robert FitzRoy gives the first public weather forecast printed in a newspaper.

1960:
NASA launches the first successful weather satellite to study our atmosphere from outer space.

2019:
The main weather-forecasting computer program used by the United States gets an upgrade, enabling it to make more detailed and accurate forecasts.

Kiribati's islands are spread out over an area of 1.4 million square miles (3.5 million square kilometers) of ocean. That's a little larger than India!

In 2014, the government of Kiribati spent $8.7 million to buy land from Fiji, another island in the Pacific, so its people could relocate there.

Warming Waters

The country of Kiribati is in the Pacific Ocean. It's made up of 32 islands and other pieces of land. More than 100,000 people live there. But now much of this land is being swallowed by rising seas. Climate change has raised global sea levels by about 8 inches (20 centimeters) since 1900. Almost half of that increase has occurred since 1993. The rising water causes problems like flooding and **erosion** for people living on coasts and islands. If the warming trend continues, these problems will get worse.

Water released by a melting glacier is called meltwater.

I'm Melting!

Where is the extra water coming from? About one-third of it comes from the melting snow and ice on mountain **glaciers**. Glaciers form in places where more snow falls in winter than melts in summer. But rising temperatures are melting glaciers faster than winter snow can build them back up. As a glacier melts, the water flows down mountains, into rivers, and out to the oceans. That raises the water level in the oceans, much like a faucet pouring water into a bathtub.

An ice sheet is a type of glacier that covers an area greater than 19,000 square miles (50,000 sq km). Today, large ice sheets in Antarctica and Greenland are melting and contributing to global sea level rise. If these ice sheets were to melt completely, sea level would rise by about 217 feet (66 meters) worldwide!

Glaciers and ice sheets store about 70 percent of the world's fresh water.

Adélie penguins live in Antarctica. Their habitat is threatened by climate change.

Soaking Up Sunlight

Losing ice has another consequence. Ice plays an important role in controlling Earth's temperature. Sunlight that hits sea ice is reflected off the surface. This sends the sun's energy back into space. But as ice melts, it uncovers more land and liquid water. These areas are darker than ice and reflect much less sunlight. They take in, or absorb, most of the sun's energy as heat. This warms the planet even more, melting even more ice.

A Dangerous Cycle

Even without melting glaciers, warming ocean temperatures would still cause sea level to rise. That's because of a phenomenon called **thermal** expansion. Most matter—including seawater—expands when it's heated. On average, the world's oceans are about 1°F (0.6°C) warmer than they were in 1880. Since warm water takes up more space than cold water, the world's oceans have expanded. Between one-third and one-half of sea level rise is due to this phenomenon.

Thermal Expansion

Molecules before heat is applied

Molecules after heat is applied

Spinning storms that form in the Atlantic are called hurricanes. Those that form in the Pacific are called typhoons.

Hurricane Harvey turned this street into a river in 2017.

Wetter Storms

In August 2017, Hurricane Harvey dumped more than 4 feet (1.2 meters) of rain on Houston, Texas. It caused massive flooding and took many lives. About 39,000 people had to evacuate their homes. One week later, Hurricane Irma pounded the Caribbean and the Florida Keys. Then in late September, Hurricane Maria devastated Puerto Rico. With 10 hurricanes, the 2017 hurricane season was the most intense ever recorded. Scientists expect climate change to make hurricanes wetter and potentially more damaging.

In 2017, Hurricane Irma hit Miami with strong winds and surging water.

How Hurricanes Form

Hurricanes are strong swirling storms that form over the Atlantic. They need two main ingredients: warm water and wind. Ocean water **evaporates** into clouds above the ocean. Wind blows on the clouds, which swirl with great strength. Today, the air and ocean are warmer because of climate change, so even more water evaporates. This can make hurricanes wetter and more damaging. In addition, the number of hurricanes each year has increased since the 1970s, though scientists are unsure if climate change is the only factor responsible.

Hurricane Strength

Hurricanes are classified by their top wind speeds.
The scale is called the Saffir-Simpson Wind Scale.

Category	Wind Speed (miles per hour)	Extent of Damage
1	74–95	Winds blow off roof shingles, house siding, and big tree branches. Power outages are possible.
2	96–110	Roofs and siding can sustain major damage, and trees might be uprooted. Power can be out for days or weeks.
3	111–129	Roofs may fly off. Trees might be uprooted and block roads. Power can be out for weeks.
4	130–156	Damage to outside walls and roofs may be severe, including loss of structures. Most trees go down. Power outages can last months.
5	157 or higher	Some houses are destroyed, while many others have severe damage. Winds can take down trees and power poles. People are not able to live in most of the affected area for weeks or months.

Climate change is expected to make hurricanes wetter, so scientists think a new scale will be needed to measure storm strength.

Rainy Days

With more water evaporation, climate change has also led to more rain. As the climate warms, scientists project that many wet regions will become even wetter. Places like the **tropics** will experience more rainfall—and heavier rainstorms. From July 2018 to June 2019, the United States experienced record rainfall. It rained an average of 37.9 inches (96.2 cm) that year. That's almost 8 inches (20 cm) more than usual!

Hurricane Harvey dumped 27 trillion gallons (102 trillion liters) of rain on Texas.

In 2014, a storm in Buffalo, New York, dumped up to 6 feet (1.8 m) of snow in 24 hours.

Weird Winters

Even though Earth's temperature is rising, it doesn't mean blizzards are a thing of the past. From 1950 to 2000, the eastern two-thirds of the continental United States got pummeled. During that period, those states experienced nearly twice as many snowstorms as in the first half of the 20th century. Although winters will likely get shorter, the same factors that lead to more rainfall can create more blizzards. Moisture in the air turns into precipitation. And if the air is below freezing, that precipitation falls as snow.

The hottest day on record was in Death Valley. The temperature hit 134.1°F (56.7°C) on July 10, 1913.

Death Valley National Park is the hottest, driest, and lowest national park in the United States.

Extreme Temperatures

Death Valley is the hottest place on the planet. The ground gets so hot that you could fry an egg on it. But Death Valley, which stretches between California and Nevada, is getting even hotter. July 2018 was the hottest month on record there. Temperatures reached 120°F (48.9°C) or higher for 21 days. That heat wave beat the previous record, which had been set just the year before.

Heat Waves

It's not just hot places that are getting hotter. In 2018, a global heat wave broke temperature records in countries farther from the equator, including Canada and Norway. And in some spots in Australia, temperatures rose above 120°F (48.9°C). In fact, the 10 hottest years ever recorded have occurred in recent years. As the planet warms, scientists project that heat waves will get longer and hotter. Periods of extreme heat will also happen more frequently.

Global Land and Water Temperature Above 20th-Century Average

Year	Amount Above Average (°F)	Amount Above Average (°C)
2009	1.17	0.65
2010	1.31	0.73
2011	1.04	0.58
2012	1.15	0.64
2013	1.22	0.68
2014	1.33	0.74
2015	1.69	0.94
2016	1.80	1.00
2017	1.64	0.91
2018	1.49	0.83

Heat and Your Body

For humans, the ideal outside temperature is about 82°F (28°C). During an extreme heat wave, people's bodies are at risk of overheating, which is dangerous. When a body overheats, the person may experience heat exhaustion first. If the condition is not treated, it may develop into heat stroke, which can be deadly. Here are the signs to look out for—and what to do.

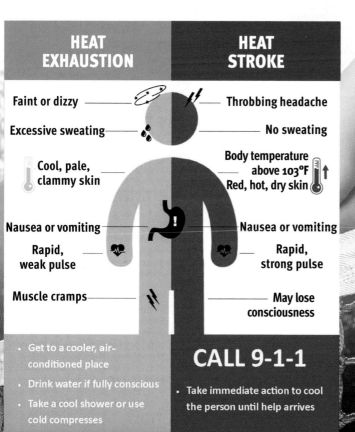

HEAT EXHAUSTION	HEAT STROKE
Faint or dizzy	Throbbing headache
Excessive sweating	No sweating
Cool, pale, clammy skin	Body temperature above 103°F Red, hot, dry skin
Nausea or vomiting	Nausea or vomiting
Rapid, weak pulse	Rapid, strong pulse
Muscle cramps	May lose consciousness
• Get to a cooler, air-conditioned place • Drink water if fully conscious • Take a cool shower or use cold compresses	CALL 9-1-1 • Take immediate action to cool the person until help arrives

Source: NOAA/National Weather Service

Sweating can cause dehydration. Drink lots of water!

When soil dries out, it hardens and cracks.

All Dried Up

While wet places will likely get wetter, the opposite is in store for dry places. Warmer temperatures in dry regions like deserts will speed up evaporation, taking moisture out of the soil. Eventually this can lead to **drought**. Drought occurs when it rains too little over a long period of time. If a drought lasts too long, it can cause crop failure and shortages of drinking water. Scientists expect severe droughts to become more likely as Earth continues to warm. This has already been happening in the Sahara in Africa. The desert is growing and swallowing up farmland in the countries of Sudan and Chad.

Rise in Wildfires

Drought can also contribute to wildfires. Soil, plants, and trees dry out when there's not enough precipitation, so they catch fire more easily. And the fire can spread more quickly through the dried areas. In recent years, major wildfires have plagued the western United States. Scientists estimate that over the past 30 years, climate change has caused twice as much land to burn as would have otherwise. The burning trees release even more greenhouse gases into the atmosphere.

Humans cause almost 90 percent of wildfires.

The Costs of Climate Change

Natural disasters can cost people their homes—and even their lives. They also cost money. During a disaster, emergency workers rush to help people. Paying these workers costs towns money. After the disaster, it's up to local, state, and national government agencies and insurance companies to help cities rebuild. Here's what some extreme events have cost in the past.

STORM DAMAGE
Hurricane Katrina

COST: $160 billion

In 2005, Hurricane Katrina pummeled New Orleans. About one million people were displaced from their homes. Many didn't return. The city's population fell by about one-third after the hurricane. Those who did return had to rebuild their homes. And because of the extensive flooding, the government had to rebuild roads, schools, hospitals, and more.

WILDFIRES
The Camp Fire

In November 2018, a wildfire that started on Camp Creek Road in California became the deadliest and most expensive in U.S. history. Firefighters and emergency response teams battled the blaze for 18 days before a rainstorm helped contain it. More than 150,000 acres (60,703 hectares) burned and more than 14,000 structures were destroyed.

COST: $16.5 billion

LANDSLIDES
Highway 1

Parts of California saw torrential rains in the winter of 2017. Rain caused a massive landslide that covered a 0.25-mile (0.4-kilometer) stretch of a busy highway. Over one million tons (907,000 metric tons) of rocks and dirt blocked the roadway. The roadway was closed for 14 months, and tourism in the area suffered during that time.

COST: $1 billion

Floods are the most common natural disaster in the United States.

In 2018, waves that reached higher than 13 feet (4 m) caused flooding in Havana, Cuba.

The Future Climate

What will the climate be like in 2100? Scientists are collecting as much data as they can about past and present climate. They use these observations and records to create computer programs that help project future climate conditions around the globe. In some potential futures, the effects of climate change are limited. In others, they are disastrous. How much things change depends a great deal on whether people limit the release of greenhouse gases today.

An Upward Trend

By the year 2100, rising temperatures might make it feel like we've all moved to different climate zones! For example, a summer day in Seattle, Washington, might feel like summer does now in Southern California. Seasons might change at different times than we are used to. Farms might have to be moved north so plants can grow. In the Northern Hemisphere, ice cover could decrease by 15 percent.

Scientists measure glacier ice to research climate change.

Lost to the Sea

Some studies project that global sea level could rise up to 6.6 feet (2 m) by 2100 if we continue to burn large amounts of fossil fuels. That would put over 691,000 square miles (1.8 million sq km) of land underwater, including cities such as Miami, Florida. Much of the Netherlands, a coastal country in Europe, would be underwater. Hundreds of islands in the South Pacific and Caribbean would disappear. About 187 million people around the world would have to move.

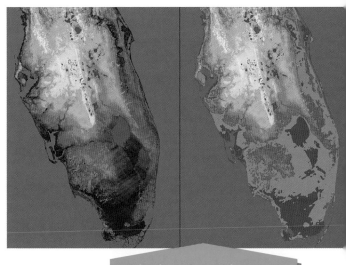

The image on the left shows Florida today. The one on the right shows what could happen to Florida if sea levels rise as predicted.

Extreme Storms

Super Typhoon Mangkhut, which hit China and the Philippines, was one of the strongest storms of 2018. The intense downpour caused deadly landslides. Scientists predict that superstorms will become more typical as the climate warms. In the United States, the Midwest and the Northeast are likely to experience more frequent downpours. However, in the Southeast, total rainfall will likely decrease, despite stronger storms.

Slowing the Change

There are ways to limit future climate change. One of the most effective actions people can take now is to stop using **fossil fuels**, which add greenhouse gases to the atmosphere. In 2016, a group of 48 countries vowed to eliminate these fuels by 2050. The group includes coastal countries and Pacific island nations that are threatened by rising seas. These countries have urged the world to join them in fighting climate change. They hope a global effort can prevent the worst effects.

Wind and solar power are energy sources that don't release greenhouse gases.

Analyzing Heat Waves

Scientists collect all sorts of information related to Earth's climate. There are several different ways to study extreme temperatures. One is to monitor heat waves. A heat wave is a period when the temperature is extremely hot for several days. Cities are hotter than green spaces because the buildings and dark asphalt absorb sunlight. The graph at right shows how the average number of heat wave days has changed by decade in 50 U.S. cities. Study the graph, and then answer the questions.

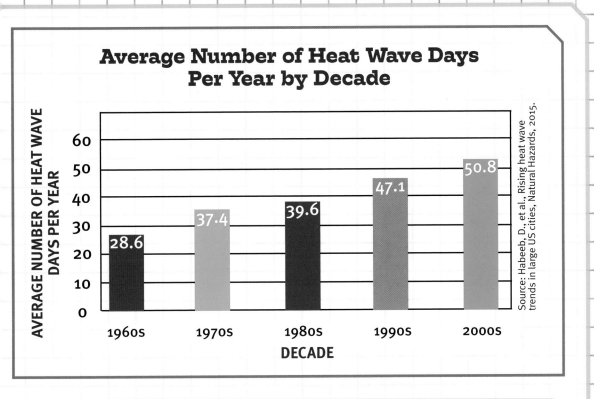

Average Number of Heat Wave Days Per Year by Decade

AVERAGE NUMBER OF HEAT WAVE DAYS PER YEAR

60
50
40
30
20
10
0

1960s — 28.6
1970s — 37.4
1980s — 39.6
1990s — 47.1
2000s — 50.8

DECADE

Source: Habeeb, D., et al., Rising heat wave trends in large US cities, Natural Hazards, 2015.

Analyze It!

1. What was the average number of heat wave days per year in the 1990s?

2. Which decade shown had the highest average number of heat wave days per year?

3. Do you see a pattern or trend in the graph? Explain your reasoning.

4. Based on this graph, would you expect more or fewer heat wave days in the 2010s than at the start of the 21st century?

CAUTION!
EXTREME
HEAT
DANGER

Hometown Heroes

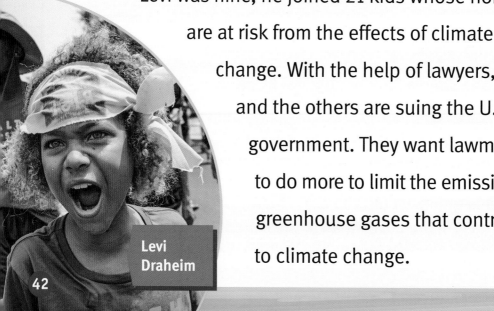

Levi
Draheim

Levi Draheim lives on an island off the east coast of Florida. He loves living near the beach, but rising sea levels threaten to permanently flood his town. When Levi was nine, he joined 21 kids whose homes are at risk from the effects of climate change. With the help of lawyers, Levi and the others are suing the U.S. government. They want lawmakers to do more to limit the emissions of greenhouse gases that contribute to climate change.

Levi speaks to the crowd outside a courthouse in Oregon.

If the kids win, the U.S. government could be forced to regulate greenhouse gas emissions. Lawsuits like theirs are just one of many ways to fight for a cause. If you want to help with Levi's cause and other problems created by climate change, keep yourself informed. Then you can write to officials in your town or state, or even members of the U.S. Congress. You can also join protests against climate change.

True Statistics

Amount average sea level has risen globally since the start of the 20th century: 8 inches (20 cm)

Projected global average sea level rise by 2100 due to climate change if emissions continue at the current rate: 6.6 feet (2 m)

Percentage of U.S. population that lives on coasts: 42

Amount of snow and ice that the world has lost since 1961: 10 trillion tons (9 trillion metric tons)

Percentage of sunlight that is reflected when it hits ice: 50; when the ice is covered by fresh snow, the percentage goes up to 90

Percentage of sunlight that is reflected when it hits forests: 11–15

Projected cost of climate change to the U.S. in 2100: $280 billion–$500 billion per year

Did you find the truth?

(F) Climate change is making hurricanes less wet than they used to be.

(T) The world's oceans are expanding because warm water takes up more space than cold water.

Resources

Other books in this series:

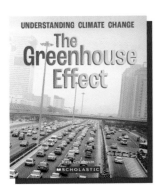  

You can also look at:

Ford, Jeanne Marie. *Extreme Weather Events*. New York: Cavendish Square Publishing, 2019.

Heos, Bridget. *It's Getting Hot in Here: The Past, Present, and Future of Climate Change*. New York: HMH Books for Young Readers, 2016.

Thomas, Keltie. *Rising Seas: Flooding, Climate Change and Our New World*. Ontario, Canada: Firefly Books, 2018.

Winchester, Simon. *When the Sky Breaks: Hurricanes, Tornadoes, and the Worst Weather in the World*. New York: Viking Books for Young Readers, 2017.

Glossary

atmosphere (AT-muhs-feer) the mixture of gases that surrounds a planet

climate change (KLYE-mit chaynj) global warming and other changes in the weather and weather patterns that are happening because of human activity

drought (drout) a long period of time without rain

erosion (i-ROH-zhuhn) the wearing away of something by water or wind

evaporates (i-vap-uh-RAY-tes) changes from a liquid into a vapor or gas

fossil fuels (FAH-suhl FYOO-uhlz) coal, oil, and natural gas, formed from the remains of prehistoric plants and animals

glaciers (GLAY-shurz) slow-moving masses of ice found in mountain valleys or polar regions. A glacier is formed when snow falls and does not melt.

global warming (GLOW-buhl WAR-ming) rise in temperature around Earth due to heat from the sun trapped by greenhouse gases in the atmosphere

greenhouse gases (GREEN-hous GAS-ez) gases such as carbon dioxide and methane that contribute to the greenhouse effect

satellites (SAT-uh-lites) spacecraft that are sent into orbit around Earth, the moon, or another heavenly body

thermal (THUR-muhl) of or having to do with heat or holding in heat

tropics (TRAH-piks) the extremely hot, rainy area of Earth near the equator

Index

Page numbers in **bold** indicate illustrations.

About the Author

Karina Hamalainen has been a writer and an editor of Scholastic's science and math magazines for a decade. Today, she is the editorial director of *Scholastic MATH*, a magazine that connects current events to the math that students are learning in middle school. She's written six nonfiction books and many articles about everything from the science of *Star Trek* to the effects of the *Deepwater Horizon* oil spill in the Gulf of Mexico. She lives in New York City.